Chocolate

chocolate, hot fudge, nuts

chocolate, hot fudge, sprinkles

chocolate, caramel, nuts

chocolate, caramel, sprinkles

Vanilla

vanilla, hot fudge, nuts

vanilla, hot fudge, sprinkles

vanilla, caramel, nuts

vanilla, caramel, sprinkles

MathStart® COMBINATIONS

The Sundae Scoop

by Stuart J. Murphy • *illustrated by* Cynthia Jabar

HarperCollins*Publishers*

To Jack—
who already knows the scoop on ice cream
—S.J.M.

To Mum with love
—C.J.

The publisher and author would like to thank teachers Patricia Chase, Phyllis Goldman, and Patrick Hopfensperger for their help in making the math in MathStart just right for kids.

For information address HarperCollins Children's Books,
a division of HarperCollins Publishers, 195 Broadway, New York, NY 10007,
or visit our website at www.mathstartbooks.com.

Bugs incorporated in the MathStart series design were painted by Jon Buller.

The Sundae Scoop

Library of Congress Cataloging-in-Publication Data
Murphy, Stuart J.
The sundae scoop / by Stuart J. Murphy
p. cm. — (MathStart)
Summary: At the picnic on the last day of school, James, his friends, and the cafeteria lady make a variety of ice-cream sundaes, using mathematics to figure out how many different kinds they can create.
ISBN 0-06-028924-4 — ISBN 0-06-028925-2 (lib. bdg.) — ISBN 0-06-446250-1 (pbk.)
1. Combinational analysis—Juvenile literature. [1. Combinations. 2. Permutations. 3. Mathematics.] I. Series.
QA164 .M87 2003 2001024322
511'.6—dc21

Typography by Elynn Cohen 23 SCP 23 ❖ First Edition

The Sundae Scoop

Winnie, the lady in charge of the cafeteria, was running the ice-cream booth for the school picnic. Lauren, James, Emily, and Winnie's cat, Marshmallow, were all on hand to help her out.

"I have a stupendous idea! Let's make sundaes!" said Winnie.

"Cool!" said James. "If we make all kinds of different sundaes, we'll have the best booth at the picnic!"

"Put on your thinking caps," said Winnie. "What kind of ice cream should we serve?"

"Chocolate!" said Lauren. "That's *my* favorite."

"Bubblegum!" said James.

"Peppermint stick!" said Emily.

"Whoa! That's too many!" complained Winnie. "Let's just have vanilla and chocolate."

"Now, what's the scoop on sauces?" asked Winnie.
"Vanilla with caramel!" said Emily. "Caramel is *my* favorite."

"Chocolate with hot fudge!" said Lauren.
"They already sound yummy!" said James.

"Fabulous!" said Winnie. "I'll draw up a chart on my chalkboard. If we have 2 kinds of ice cream and 2 kinds of sauce, that makes—let's see—how many kinds of sundaes?"

"It looks like 4," said James.

SUNDAE COMBOS
Vanilla
HOT FUDGE
CARAMEL
Chocolate
HOT FUDGE
CARAMEL

“What about nuts?” said Lauren.

“And sprinkles?” said James. “Sprinkles are *my* favorite.”

“Right on,” said Winnie. “Sundaes need toppings, too.”

“I hope that will be enough different combinations,” said Emily, frowning.

“It’s more than you think,” said Winnie.

"Each sundae will have one flavor of ice cream, one sauce, and one topping," said Winnie. "The first combination is vanilla, hot fudge, and sprinkles."

"Oh, I get it," said Lauren. "Or you could have vanilla, caramel, and sprinkles."

"That's great!" said James. "Now there are 8 different choices. That's plenty."

Vanilla
HOT FUDGE
CARAMEL
Sprinkles
Nuts
Sprinkles
Nuts
Chocolate
HOT FUDGE
CARAMEL
Sprinkles
Nuts
Sprinkles
Nuts

The day of the picnic was sunny and warm. Everybody wanted sundaes.

"Let's get scooping!" said Winnie.

"Look at the line," Lauren whispered to Emily.

Emily looked up. "I hope we still have all *our* favorites left for ourselves," she said.

ice-
cream
sundaES
vanilla
Chocolate
Hot
Fudge
CARAMEL
NUTS

Emily scooped. James poured the sauce. Winnie added the nuts. And Lauren did a little dance as she shook out the sprinkles.

"And a one . . . and a two . . .
and a one, two . . . WHOOPS!"
she said.

"That was all the sprinkles we had!" complained Emily.
"Yeah," said James. "There goes *my* favorite."
Marshmallow didn't seem to mind.

“Better change the sign,” said Winnie.
“Now we’re down to 4 kinds of sundaes.”

James started to pour the caramel sauce for the next sundae.

"Watch out for Marshmallow," said Lauren.

"Where?" asked James.
"James!" said Emily. "Look where you're pouring!"
"Oops," said James.

"That's it for the caramel sauce," said Lauren.

"*My* favorite," said Emily sadly.

"Now there are only 2 sundaes to choose from," said James.

Vanilla
HOT FUDGE
CARAMEL
Sprinkles
Nuts
Sprinkles
Nuts
Chocolate
HOT FUDGE
CARAMEL
Sprinkles
Nuts
Sprinkles
Nuts

The sun got hotter and hotter. "Scoop faster!" Winnie told Emily. "The chocolate ice cream is turning into chocolate soup!"

Emily scooped as fast as she could. But it wasn't fast enough.

“There goes *my* favorite,” sighed Lauren.

“It looks like it’s Marshmallow’s favorite too,” said James.

“Better change the sign,” Winnie said.

"That was the last person in line," said Emily. "Thank goodness! Now we can have our *own* sundaes."

"But there are no more sprinkles," said James.

"Or caramel sauce," said Emily.

"No chocolate ice cream either," added Lauren.

"Meow," said Marshmallow.

"By gum, you're right!" Winnie said. "There's just one kind of sundae left."

"Vanilla ice cream, hot fudge, and nuts," said Winnie.
"*My* favorite! Pass me a spoon!"

FOR ADULTS AND KIDS

In *The Sundae Scoop*, the math concept is combinations. Determining how many different combinations can be made from a given group of items is an important problem-solving and pre-algebra skill.

If you would like to have more fun with the math concepts presented in *The Sundae Scoop*, here are a few suggestions:

- Read the story with your child and describe the diagrams that show the number of sundaes. Discuss how the diagram changes as the situation in the story changes.
- As you reread the story, ask questions like, "How many flavors of ice cream are there? How many different sauces? How many toppings? How many different sundaes could the kids make?"
- Re-create your own sundae scoop story. Have your child think of several different flavors of ice cream, sauces, and toppings, and write them down. Help your child draw diagrams similar to those in the story to determine the number of different sundaes they could create with their imaginary ingredients.
- Make up combinations stories with your child. For example, say, "The school store sells red and blue pencils. The pencils have pink, blue, or green erasers. You can have your name printed on a pencil with green or yellow ink. How many choices of pencils do you have?" Help the child draw diagrams like the ones in the story to find out.

Following are some activities that will help you extend the concepts presented in *The Sundae Scoop* into a child's everyday life:

Ordering Lunch: Using a fast-food restaurant menu, have your child choose his or her favorite sandwich. Then help your child pick 2 favorite drinks and 3 favorite desserts. Determine how many different lunches could be served using these items.

Decorating Cookies: Bake batches of 2 different types of cookies, such as sugar cookies or oatmeal cookies. Have 2 colors of icing and 3 types of sprinkles available. Help your child figure out how many different combinations there can be if each cookie is decorated with 1 color of icing and 1 type of sprinkles.

Getting Dressed: Lay out 2 pairs of shoes, 4 shirts, and 2 pairs of pants for the child. Help your child determine the number of different outfits that he or she could wear.

The following stories include concepts similar to those that are presented in *The Sundae Scoop*:

- A Three Hat Day by Laura Geringer
- Each Orange Had 8 Slices by Paul Giganti Jr.
- How Hungry Are You? by Donna Jo Napoli and Richard Tchen

Chocolate

chocolate, hot fudge, nuts

chocolate, hot fudge, sprinkles

chocolate, caramel, nuts

chocolate, caramel, sprinkles

Vanilla

vanilla, hot fudge, nuts

vanilla, hot fudge, sprinkles

vanilla, caramel, nuts

vanilla, caramel, sprinkles